Astronomy
General Science Lab Manual

Written and compiled by
Christopher Claysmith

Astronomy: General Science Lab Manual

ISBN: 978-1-943536-64-1

Chemeketa Press

Chemeketa Press is a nonprofit textbook publisher that works directly with faculty authors to make affordable, effective, engaging, and accessible textbooks. We bring the passion and enthusiasm of your favorite professor to the page, through agile publishing methods that will change the industry. We do this for our students and yours, nationwide—because a textbook should open the door, not close it.

To learn more, visit www.chemeketapress.org

Publisher: David Hallett
Director: Steve Richardson
Managing Editor: Brian Mosher
Design Manager: Ronald Cox IV
Cover Design: Brice Spreadbury
Interior Design and Layout: Matthew Sanchez
Cover: he cover image is a compilation by Brice Spreadbury of several images in the public domain.

Printed in the United States of America.

Contents

Lab 1a: **What's in the Bag?**

Purpose

In this lab activity, your goal is to determine the contents of the black bag at your station. In addition, you will draw a detailed diagram of how the contents are placed within the bag. Along the way, you will gain some insight and experience with the scientific method.

Background

You will mainly be dealing with the scientific method and terms related with the models (theories). Some of the terms you should become more familiar with include: hypothesis, theory, conclusion, observation, and inference.

Instructions

You will come up with much of the procedure for this experiment on your own.

1. To get you started, do the following. Without talking to your lab partner, come up with a definition for the terms **observation** and **inference** in your own words and write them in the space below:

2. Now, discuss your definitions as a lab team. The goal is to come up with definitions of the terms that you are both comfortable with. After your discussion, write the definitions for the two terms in the space below:

3. Next, you should compare your definitions to those found elsewhere to see how consistent you are. Start the computer at your station, open a Web Browser, go to a search engine (such as Yahoo, Excite, Google, etc.), and search the web for definitions of the terms *observation* and *inference*. Be careful with your search, however, as information contained on a web page does not necessarily go through any sort of a review process like a book does to become published. This means you have to be careful of any agendas a particular website has when using information from that site. Discuss the various definitions that you find with your lab partner, and decide on the ones from the web you think are the best and write them in the space below. (Also, write down the web address for the page where you found the definitions.) Let your lab instructor know when you are at this point so that your "black box" can be delivered to your station.

Data

1. Once the bag is set down at your station, one person in the group should give the bag a gentle shake. The person should then set the bag down and the two of you should come up with your *hypothesis* for what is in the bag. Write it in the space below, along with your reasons for the hypothesis. After each reason, place an (O) if it was an observation or an (I) if it was an inference. Make sure that you discuss each with your lab partner and that you agree before writing them down.

2. Now that you have a hypothesis, you should begin to devise ways to test to see if the hypothesis is correct. (Almost every hypothesis must be modified at some point, if it is not completely thrown out. Do not be discouraged if you wind up substantially changing your hypothesis during the course of the lab. Remember that the hypothesis is a starting point.) As you conduct your tests, record the information you determine from each test in the space below. Again, after each piece of information place an (O) if the information was determined by observation and an (I) if it was determined by inference. Continue this process until you believe you know for sure what is in the bag.

Conclusions

1. Start by writing what you believe to be the contents of the bag in the space below.

2. Now draw a diagram of how you believe the contents are placed in the bag.

3. Explain why you feel the above is correct.

4. In a few sentences, summarize what you feel are the important ideas that you gained from this lab activity. Also, list any questions you still have after having completed the activity (besides, "what is in the bag?").

Lab 1b: **Welcome to Stellarium**

Purpose

An introduction to the Astronomy tool Stellarium.

Materials

Stellarium can be downloaded for free at: www.stellarium.org

Background

Stellarium is an open source and free-to-download planetarium program that we will be using during many GS107 labs. It is a powerful tool for visualizing the night sky and seeing how objects in the sky move throughout the day, year, and millennia. The table below describes the operations of buttons on the main tool-bar and the side tool-bar, and gives their keyboard shortcuts.

TIP: If you aren't sure what one of the buttons on the control bar does, hover the mouse pointer over it for more information.

Feature	Tool-bar button	Key shortcut	Action
Constellations		c	Draws constellation lines
Constellation Names		v	Draws the name of constellations
Constellation Art		r	Superimposes artistic representations of constellations over the stars
Equatorial Grid		e	Draws grid lines for the RA/Dec coordinate system
Azimuth Grid		z	Draws grid lines for the Alt/Azi coordinate system
Toggle Ground		g	Toggles drawing of the ground. Turn this off to see objects that are below the horizon
Toggle Cardinal Points		q	Toggles North, South, East and West points on the horizon

Toggle Atmosphere		a	Toggles atmospheric effects (note: makes the stars visible in the daytime)
Nebulae & Galaxies		n	Toggles positions of Nebulae and Galaxies, even when FOV is too wide to see them
Planet Hints		p	Toggles indicators for position of planets
Coordinate System		Enter	Toggles between Alt/Azi & RA/Dec coordinate systems
Goto		Space	Centers view on selected object
Night Mode		[none]	Toggles "night mode," changes color of some display elements to be easier on the dark-adapted eye
Nebula background images		[none]	Toggles "nebula background images," turns textures on or off
Full Screen Mode		F11	Toggles full screen mode
Flip image (horizontal)		CTRL+SHIFT+h	Flips image in the horizontal plane (note: button is not enabled by default, see section [sec: imageflipping])
Flip image (vertical)		CTRL+SHIFT+v	Flips image in the vertical plane (note: button is not enabled by default, see section [sec: imageflipping])
Quit Stellarium		CTRL-Q	Closes Stellarium (note: keyboard shortcut is COMMAND-Q on OSX machines)
Help Window		F1	Shows help window, lists key bindings and other useful information
Configuration Window		F2	Shows display of the configuration window
Search Window		F3 or CTRL+f	Shows display of the object search window
View Window		F4	Shows view window

Time Window	⊘	F5	Shows display of the help window
Location Window	✳	F6	Shows observer location window (map)

Part 1: Navigating the Celestial Sky

1. In Stellarium, the first thing you need to do is set your location. On the left hand bar, click the **Location Window (F6)** to open a search bar. Search for Salem, Oregon, and be sure to set it as the default location. You can set any position in the world (including your own backyard if you know the Latitude and Longitude) later on.

2. Spend some time getting used to the Stellarium interface. Try all the buttons on the bottom bar to get used to what they do. The most important ones will be **Ground (g)** and **Atmosphere (a)** since those get in the way of seeing objects. For now, turn off the **Atmosphere** only. How does the view change when the atmosphere is off?

Open the **Sky and viewing options window (F4)**. Click on **Markings**. Try turning on the Equatorial Grid (J2000). Click Fast Forward to let the stars move some. Turn off the Equatorial Grid and turn on the Azimuthal Grid. Click Fast Forward to let the stars move some more. Which grid is mapped to the stars? Which grid is mapped to the Earth?

3. What situations would each grid be useful for?

4. Turn on **Constellations Labels** and **Constellation Lines**. If you are not sure which Zodiac sign you are, look it up online. What do you think the Zodiac has to do with your birthday?

5. Let's test your theory. Set the Stellarium date to your birthday (the exact time doesn't matter for us, but astrologers would disagree). Where is your Zodiac constellation? Where is the Sun?

6. Turn on the Equator (J2000) and the Ecliptic (J2000). Find the location on the sky where the two lines cross. Using the clock, find the date when the Sun is at that position. When is the Sun on the Celestial Equator? What do we call this date? (You can click on the Sun and look at the RA/Dec. The Sun is on the Celestial equator when the Dec is 0 degrees. Press ALT and either the – or = keys to keep the sky in the same position but move the Sun and planets)

7. Set the date so that the Sun is at its highest above the celestial equator. What do we call this event? Do the same with the date when the Sun is lowest below the equator. What do we call this event? (The Sun is highest when the Dec is highest and lowest when the Dec is lowest)

8. Go to stars.chromeexperiments.com (Or just search in Google for "100,000 Stars"). This is a 3D map of the 100,000 stars closest to the Sun. First, zoom into the Sun and the Solar System to get an idea of the distance scales, and then zoom out until you can see the names of the closest stars. Choose five of them, then guess if they would appear bright or dim.

9. Now look up your five stars in Stellarium. Find the apparent magnitude of each star and write it down. This is a way of comparing how bright stars appear in the night sky. The lower the apparent magnitude, the brighter the star appears in the sky. If you can't find the star in Stellarium, assume that it is above a magnitude 12 star.

10. Are these nearby stars bright or dim? A bright star is magnitude 1 to 3, dim is 4 to 6, and too dim to be seen by naked eye is greater than 6.

11. Why do think that nearby stars appear the way they do? What are the two factors that affect the apparent brightness of a star?

12. Find the North Star, Polaris. Where is it on the equatorial grid? How about on the azimuthal grid?

13. Set the year to 100 years ago, 500 years ago, 1000 years ago, and 10,000 years ago (you can put in negative dates for BCE) What is happening to Polaris? What are some reasons why this may happen?

Part 2: Using Stellarium and the Internet

1. Write down a question you've always had about Astronomy. Then write down a hypothesis that you think would be the answer to your question.

2. Use Google and Stellarium to find the answer. Was it close to what you thought? Why or why not?

3. Find the Moon and zoom in on it. Do you see the "Man in the Moon?" Or are you like some people and instead see the Rabbit in the Moon?

4. Set the location of Stellarium to Sydney, Australia. How does its night sky compare with Salem's? Is there anything different about the Moon?

5. Find information on when the next comet will be visible and when the next solar and lunar eclipses will be. Write down the dates you found below.

6. What is the Philae spacecraft? What did it do and what went wrong?

7. Pick one of the following and find an artist's representation: Neutron Star, black hole, exoplanet, or Milky Way Galaxy. Write down your pick.

8. Describe what your chosen object looks like. What colors are there? What details make it stand out in comparison to planets and stars?

9. Now find an observation of the same thing. How does it compare? Be specific.

10. Why do observations and artist's representations differ?

11. Why would a scientist use an artist's representation?

Lab 2: **Motion of the Sun, Moon, and Planets**

Purpose

To examine the motions of the Sun, Moon, and planets in Stellarium.

Background

As we have discussed in the last lab, there are many different ways we can map the night sky. If we use the equatorial grid, stars are in locked positions as the grid "rotates" around the Earth. However, the Sun, Moon, and the planets aren't nearly as stationary.

Part 1: The Stars and the Celestial Sphere

1. Click on **Sky and Viewing Options** and click on **Markings.** Turn on the **Equatorial Grid (J2000)** and the **Meridian**. Choose a bright star (Not a planet!), and click on it to target it. Set the time of day so that your star is positioned on the meridian. Record the Az./Alt. position and the time. You will want to pause the simulation by clicking on the Play button to stop the stars from moving. Be precise! Minutes and seconds matter!

2. Jump ahead in time by one day. Your star should have moved slightly. Carefully adjust the time so the star is back on the meridian. Record the time with minutes and seconds again.

3. How much time does it take for the Celestial Sphere to go around the Earth completely?

4.   Choose a bright star. Adjust the time of day until the star is directly on the Eastern Horizon. It's "on" the Eastern Horizon when it is just visible and has an altitude angle as close to zero as possible. You want the star to be directly on the horizon, not above it. At what time is it located on the Eastern Horizon?

5.   Use the **Date and Time control** to go to next month. What time is the star on the Eastern Horizon now?

6.   Go forward another month. Is the change in time constant? Why do stars rise at a slightly earlier time every night?

7.   Find Polaris. Is there any time of the day or year when Polaris is below the horizon? Why or why not?

8.   How does the Big Dipper move relative to Polaris over a 24-hour period? Could you potentially use this as a clock? Why or why not?

Part 2: The Sun

1. Locate the Sun. Click on it to display information about it. How far away is it from the Earth?

2. Using the internet, look up the dates for the vernal equinox, summer solstice, autumnal equinox, and winter solstice for this year. What defines these dates?

3. Set the Stellarium date to the vernal equinox. Find local Noon by turning the viewpoint due south and seeing when the Sun is at its highest point on the **meridian**. You can use the **Azimuthal grid** to help with this. What is the altitude angle above the horizon? What is the distance to the Sun?

4. Find the angle above the horizon for the solstices and autumnal equinox. What is the distance to the sun for each of these dates? Does the distance to the sun affect the weather on Earth? If not, what does?

5. Based on observations of the Sun and background stars or constellations in Stellarium, draw a rough sketch of the relative locations of the Earth, Sun, and background stars or constellations from the point of view of an observer outside the solar system. Make this sketch consistent with the observations at each date. Do not simply draw what you see in Stellarium.

6. Look up the term "Manhattenhenge" or "Chicagohenge." Why is this significant to people in New York City or Chicago? Can you think of any other significant alignments with the Sun that people pay attention to?

7. The Zodiac is based on which constellations the Sun is in at any given day. To illustrate how difficult an observation this is to make, set the time to noon and turn the sky on. Can you observe what constellation the Sun is in? How would you go about working out current zodiac constellation without Stellarium?

8.	Set the year to –10000, –5000, 1, and 1000. Do the stars seem all that different? What changes? Look around the entire sky to check.

Part 3: The Moon

1.	Turn off the atmosphere (a) and the ground (g).

2.	Find the Moon and click on it. Lock the view on the Moon by pressing the spacebar when it is selected.

3.	Use the **Date and Time bar** to step forward one day at a time. (You can use the fast forward button instead if you don't get motion sick very easily. You can press the = key next to Backspace to advance the time by 24 hours) What do you notice about the phase of the Moon?

4.	There are four major moon phases: New Moon, First Quarter, Full Moon, and Third Quarter. Zoom in on the Moon to see the phase. Find when the next Full Moon will be and record the date and time when it is on the **Meridian**.

5.	Skip ahead in time until the Moon is in the Third Quarter phase. Adjust the time so that the Third Quarter Moon is on the Meridian. What time of day does this happen?

6.	Find the date and time for the New Moon on the Meridian and the First Quarter Moon as well.

7. How long is one lunar cycle? (the time from one full Moon to the next)

8. Do we ever see the "dark side" of the Moon? Is it always dark?

9. Sketch the relative position of the Moon, Earth, and Sun for the New Moon, First Quarter, Full Moon, and Third Quarter. You will need to adjust the time of day to place the Moon on the Meridian. Which of these positions would allow for a Lunar Eclipse? A Solar Eclipse?

10. Search the internet for the "far side of the Moon." How does it compare with the face that we get to see?

11. Discuss with a partner or group: "Does the Moon rotate about a central axis similar to the Earth?" Write down your conclusions.

12. Compare that with the simulation from the following internet page. You need to use Internet Explorer to do this, as it won't work in Chrome. http://astro.unl.edu/classaction/animations/lunarcycles/lunarapplet.html

13. How do the phases that you can see depend on the relative locations of you, the Sun, and the Moon?

Part 4: Motion of the Planets

1. Turn on the **Ecliptic** in **Sky and Markings**. The ecliptic is the plane of the sky that the sun and the planets are always found on. How does the altitude angle for the ecliptic change over the course of a year? (check each month at the same time of day)

2. Set the date to December of 2013. Turn off the ground and find Mars. Select it, then press Spacebar to keep it centered in the screen.

3. Advance the time by one **sidereal day** by holding down **ALT** and **=** at the same time. What direction (Left or Right) is Mars moving relative to the background stars?

4. Keep going through March. What direction does it move relative to the background stars? Record the date when Mars begins to move "backwards."

1. When does Mars go back to moving the way it did in December? Record this date as well.

2. Using your answers for 5 and 6, how many days does Mars spend in retrograde?

3. Search the internet for the next time Jupiter will be in retrograde and examine its motion. Does it stay in retrograde for a longer or shorter time than Mars? Why would this be?

Lab 3: **Planetarium and Astronomy Terms**

Purpose

For this lab, we will be spending time discussing the night sky and some common stargazing techniques and terms while in the planetarium. As we go along, there will be pauses where the lights are lifted slightly, during which you should write definitions for all the following terms in your own words.

Definitions

1. Constellations, Stars, and Planets

2. Orion (Constellation)

3. Sirius (Brightest Star in night sky)

4.   Milky Way (Our galaxy, looking away from the center of the galaxy)

5.   Big Dipper (Asterism, part of Ursa Major constellation)

6.   Polaris (North Star)

7.   Leo (Constellation)

8. Regulus (Brightest star in Leo)

9. Virgo (Constellation)

10. Spica (Brightest star in Virgo)

11. Jupiter (Planet currently near Virgo)

12. Saturn

13. Mars

14. Summer Triangle (Asterism of Deneb, Vega, and Altair)

15. Milky Way (Our galaxy, looking toward the center of the galaxy)

Name _____ Section _____ Date _____

Group Members _____

Lab 4: **Sunspots and Solar Activity**

Purpose

To observe solar activity using modern astronomy tools and spacecraft.

Materials

A computer with internet access.

Acknowledgments

Based on the SDO lab by Scott Hildreth, Shannon Lee, and Timothy Dave of Chabot College.

Background

There are many stars that dot the night sky, but the most important star is the one that you can only see in the day: the Sun. Nearly every organism on Earth depends on the Sun for survival. While it is necessary for our existence, the Sun may also be our undoing. Our modern way of life depends on technology, and the more dependent we become the more vulnerable we are to the Sun's outbursts.

On March 13, 1989, an estimated 6 million people in the region of Quebec Canada lost power for 9 hours. This large blackout was caused by the Sun. A massive eruption called a *coronal mass ejection* had occurred on the Sun and was aimed right at Earth. These eruptions are actually quite common and we call this connection between solar activity and its impact on the Earth *"Space Weather"*.

The **Solar Dynamic Observatory** (SDO) is a space telescope that was launched on February 11, 2010. SDO is sitting in a geosynchronous orbit about 22,000 miles above the surface of the Earth. There are three instruments aboard: Atmospheric Imaging Assemble (AIA), Helioseismic and Magnetic Imager (HMI), and Extreme-Ultraviolet Variability Experiment (EVE).

For this lab, we will be using real data from AIA and HMI instruments to observe sunspots and solar eruptions. These observations will be reported to Stanford to be used in current research in NASA's "Living with a Star" investigations. You can read more about SDO at the NASA website: http://sdo.gsfc.nasa.gov/mission/about.php

Part I: Solar Observation Movies and Images

There are some fantastic collections of data from the instruments aboard the SDO spacecraft that have been made public on the web. Let's look at some examples.

Video 1: http://www.youtube.com/watch?v=U_MKL_fjDLo&feature=youtu.be

This is a compilation of data from both the AIA and HMI instruments for the first year that SDO was in operation. **Prominences** are solar eruptions that are fairly stable. They last for hours or even days as charged particles of gas flow along **magnetic field loops**. Bright glowing gas will look like water flowing from a hose into the Suns atmosphere. **Flares** are seen as bright bursts of energy that last for only a brief moment. You can also see **sunspots** as they evolve over time as they cross the solar surface. Sunspots look like very dark patches on the surface, almost like solar acne. Notice that the Sun is imaged in different colors. Each color represents a different kind of light and temperature.

Watch the video carefully and try to identify which parts of the video are showcasing the different types of solar activity. Below, write down the time on the video when each feature appears.

1. **Prominences:** video times _____,_____,_____,_____,_____

2. **Solar flares:** video times _____,_____,_____,_____,_____

3. **Sunspots:** video times _____,_____,_____,_____,_____

4. **Unknown features:** video times _____,_____,_____,_____,_____

Now go to the following website http://sdo.gsfc.nasa.gov/data. Here we can see the data from the Sun today. Each image has been labeled with important information. At the top, you will see "AIA" and then a number (like "AIA 193" in Figure 1). This tells you that the AIA instrument collected the data and that the wavelength of light is 193 **Angstroms** (or).

1. Compare and contrast the different wavelength images. Specifically comment below on which of the images seem the most "active" and which seem the most "quiet" and describe the criteria you are using to differentiate quiet from active.

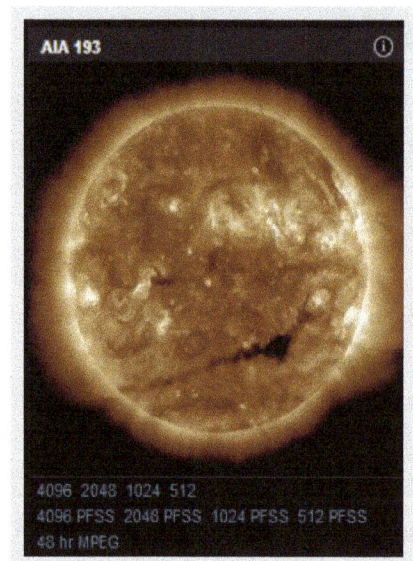

Figure 4.1

2. For any of the AIA wavelength channels, click on each of the numbers (4096 2048 1024 512) directly underneath the image. Each is a hyperlink. What is the difference between the numbers? Try zooming the picture in or out if you can't tell very easily.

3. For any of the AIA wavelength channels, click again on each of the numbers labeled with "PFSS" underneath the image. PFSS stands for Potential Field Source Surface model, and the lines you see are approximated magnetic field lines. What do you notice about the number of lines and the "density" of lines (how many are packed into a small region) that you see coming from particular features on the Sun?

Note: You can read more about PFSS models and these lines at:

Young, C. A. (2010, October 10). Magnetic field lines galore (SDO pick of the week). *The Sun Today*. Retrieved from www.thesuntoday.org/current-observations/magnetic-field-lines-galore-sdo-pick-of-the-week.

4. Each AIA wavelength channel represents a different region and temperature on the Sun. Let's explore that in more detail. The Wikipedia page for the SDO lists the wavelengths, regions, and temperatures (https://en.wikipedia.org/wiki/Solar_Dynamics_Observatory). What region of the Sun and temperature do images taken with wavelength 193 A represent?

Video 2: http://www.youtube.com/watch?v=BVDsQkuRt-c&feature=youtu.be

An *active region* is an area on the Sun where the magnetic fields are stronger than the surrounding area. *Sunspots* are visible manifestations of active regions. Active regions are also responsible for solar eruptions like *flares* and *prominences*. In this video, we see an active region in three different wavelengths of light side by side.

Figure 2 is an artistic rendering of an active region. The area under the active region can be represented by a bar magnet with a north and south pole. Charged particles are trapped along the magnetic field lines and create unstable glowing loops above the active region.

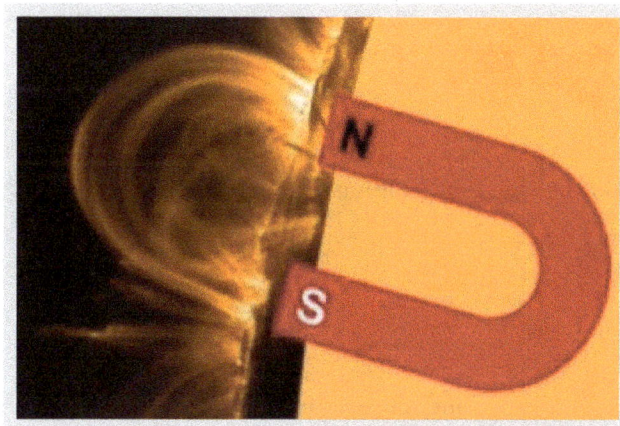

Figure 4.2: Windows to the Universe original artwork by Randy Russell using an image from NASA's TRACE spacecraft.

1. What features of the Sun are *uniquely* visible in each of the three frames of the video?

    a. Orange (AIA 304 A, Chromosphere, 50,000 K)

    b. Yellow (AIA 171 A, Corona, 630,000 K)

    c. Black and White (HMI intensitygram, Photosphere, 5,000 K)

Video 3: http://www.youtube.com/watch?v=oInoNnPsxcA&feature=youtu.be

Now let's focus our attention on sunspots. This video is one solar rotation worth of images in white light (photosphere: 5,000 K). The Earth rotates on its axis once every 24 hours (one day). The Sun also rotates on its axis but it takes much longer than one Earth day.

1. How long is one solar rotation (in Earth days)?

2. How many active regions can you count on the Sun in this solar rotation?

3. Do you think this is a time of high solar activity or low solar activity? Explain why.

4. Are there any active regions that you believe would be more likely to cause a solar eruption than others? Explain!

5. How could you check this hypothesis?

Video 4: http://www.youtube.com/watch?v=Nnwqkm6rL4M&feature=youtu.be

Active regions on the Sun are carefully labeled and catalogued. Each sunspot is given an active region number. This may seem a simple categorizing task but as sunspots evolve they can shrink, grow, split or disappear altogether. Video 4 is data from the same dates as video 3 (but only a few days instead of the full solar rotation). Here the active regions are labeled and we can focus on one in particular, region AR1339 on 11/06/11. The video will show you many details about this active region.

1. Pause the video at the **5 second** mark. How large is the sunspot group, in terms of our planet's diameter? (In other words, how many "Earth's" **across** is this sunspot group?)

2. How many Earth's **tall** is this sunspot group?

3. Area is calculated by multiplying length times width. Multiply your answers from the previous two questions. Approximately how many Earth's in area is AR 1339?

4. The next portion of the video deals with magnetic fields. Sunspots are really pairs of magnetic poles, when you get a large group of sunspots it can be very hard to separate them. Draw a quick sketch of this image and label the N and S polarity of the main magnetic fields.

5. The last portion of the video looks at sunspot evolution data. Watch carefully and see if the sunspots rotate. Is it easier to see the large spots or the small spots rotate?

6. Choose one or two spots to watch and draw them below. Add arrows to indicate which way the sunspots are rotating.

Part 2: Make your own

Now that we have some background knowledge of the Sun we can start to explore ways to make movies of our own using the same data! First we need to find an "active day" on the Sun as a starting place. We already know from the previous YouTube videos that November 2011 was particularly active (with lots of large sunspots). But some more recent dates have been very interesting, too.

We will go to the online database for the Lockheed Martin Solar and Astrophysics Laboratory (LMSAL) and search their Heliophysics Events Knowledgebase (HEK). The database can be found at http://www.lmsal.com/isolsearch.

We need to restrict ourselves to just a few hours of the Solar data to see the results clearly. First, enter **2012-10-23T00:00:00** in the *start date* box and **2012-10-23T04:00:00** in the *end date* box. Next, restrict the visible event types to only **active regions** and **flares**. To do this quickly, uncheck the preselected boxes for all event types on the left side of your screen, then check just the boxes for Active Regions and Flares. Your screen should now look like this:

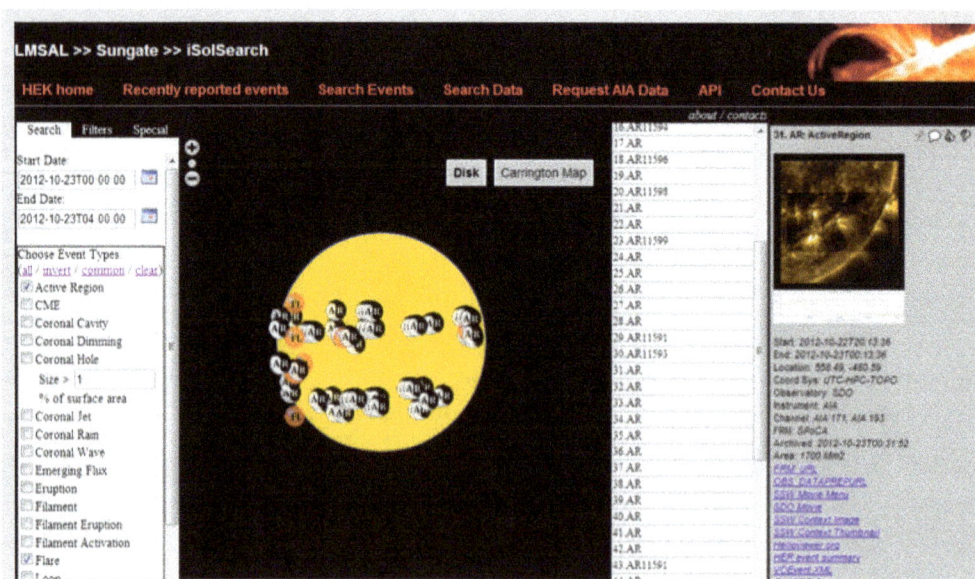

Figure 4.3

To the right, you have a list of search results. If you mouse over the spots on the Sun you should see the particular event (region) highlighted on the right. If you click on them you can get some links to appear on the far right, including movies. Some active region results will have more useful data than others.

Complete the following activity two times using the fields provided below.

1. If you click on result "**XX. AR**" (where XX is a number), you will see the data menu pictured above on the right. Click on the picture at the top, which will have the actual active region number labeled (starting with 15XX).

2. Next, click on **SSW Movie Menu** for an active region (e.g., 12.AR). You will see a Movie link that will download a short movie of the region.

3. Now, Pick a few **active regions** between results "12.AR to 18.AR" and "52.AR to 57.AR". Click on the SSW Movie Menu, and view the Flash or JavaScript representations of the event. Record your observations briefly.

4. Finally, with your team, develop at least one research question to explore further based upon the data you have seen. See an example below.

Observation example

SDO Recording of observations for (Date/Time): **2012-10-23T00:00:00-04:00:00**

Active Region # **31**

Data Viewed (JavaScript Movie, Flash Movie, etc.) **Flash Movie**

Observations: *(In this area, write about anything you noticed. Did you notice: patterns, loops, whorls, structures? Or events like a flare or sudden brightening of the image? Try to be as specific as possible about what you see and when it occurred.)*

Visible active region with loops of glowing gas flowing from one sunspot to another. A small amount of brightening on the left.

Observation 1

SDO Recording of observations for (Date/Time): _____

Active Region # _____

Data Viewed (JavaScript Movie, Flash Movie, etc.): _____

Observations:

Observation 2

SDO Recording of observations for (Date/Time): _____

Active Region # _____

Data Viewed (JavaScript Movie, Flash Movie, etc.): _____

Observations:

Review

1. From this exercise, what are the some of the most significant things you have learned about the Sun?

2. From this exercise, what are the some of the most significant things you have learned about the process of data analysis that Solar Astronomers go through?

3. What would have liked to do with this activity that you were not able to do, or not given enough time to do?

4. What would you keep, and what would you change overall, in this lab activity?

Lab 5: **Spectral Analysis**

Purpose

To examine electron orbital structures. Every substance has a unique electron orbital structure. Consequently, each substance creates a unique pattern of light (spectrum) and can be identified using this spectrum.

Materials

▷ Spectroscope

▷ Spectral glasses

▷ Colored pencils

▷ Computer

Background

When white light from an incandescent lamp is passed through a prism, it produces a continuous spectrum of colors. We see this often in Oregon in the form of the rainbow that is formed when sunlight passes through a matrix of raindrops. The different colors of light represent different wavelengths. Blue light has a shorter wavelength than red light. Pure white light is a combination of all the various colors.

If the light from a gas discharge tube that contains a particular element is passed through a prism, only narrow colored lines are observed, as opposed the continuous spectrum in a rainbow. Each line corresponds to light of a particular wavelength. The pattern of lines emitted by an element is called its line spectrum, and each element has its own characteristic spectrum.

Physicist Niels Bohr put forth an explanation of line spectra (plural of spectrum) in 1913. He made the suggestion that electrons are quantized, meaning that they cannot have just any amount of energy but can have only certain specified amounts. The specified energy values for an electron are called its energy levels. The higher the energy level of an electron, the higher its orbital pattern around the atom.

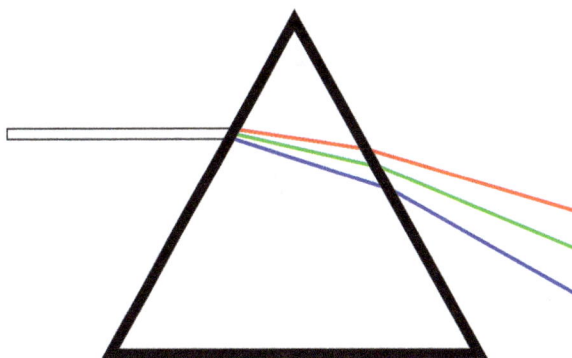

Figure 5.1

An electron can be bumped up to a higher orbital by absorbing a quantum of energy — one way to do this is to heat the atom. After a while of orbiting at a higher level, the electron will spontaneously fall back to a lower energy level. In doing this, it is going to have to give up a unit of energy in the form of a photon.

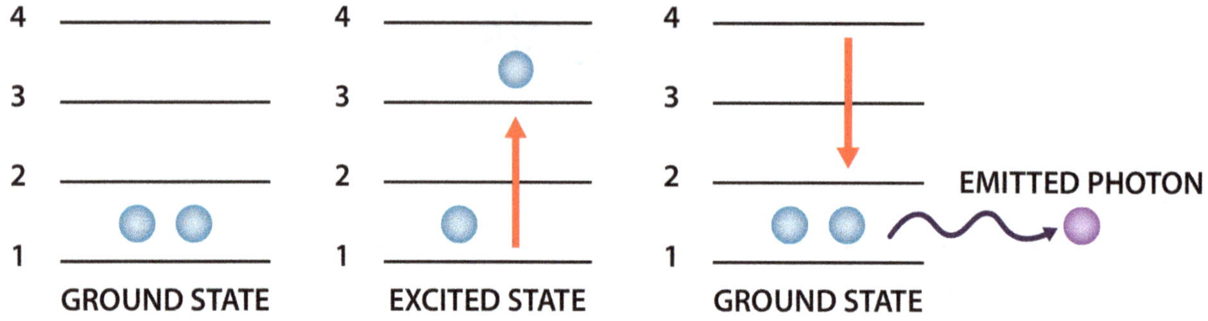

Figure 5.2

Every atom of a specific element always has its electron orbitals in the same place. In the modeled atom above, every time an electron falls from the second excited state to the ground state, a particular frequency of light will result.

Because every element has a different electron orbital structure, we should be able to identify a substance by the characteristic spectral lines that are emitted when we expose it to high heat or electrical current.

Introductory Questions

1. Think about lights you have seen. List several different types of lights that have different colors. *ex: street lights – slightly orange*

2. If you could spread out the light from all of the sources (in question 1) to see their spectra, would you expect the spectra to all look the same?

3. On a computer, go to the following web page and click on the simulation to download it. https://phet. colorado.edu/en/simulation/legacy/discharge-lamps. Be sure to turn on the spectrometer. What kinds of light do you get from hydrogen? Now try neon, sodium, and mercury. What kinds of light do you get?

4. Why are only certain colors emitted? Think about the energy level diagrams on the right-hand side of the screen.

Excitation Spectra

During today's lab, a tube of hydrogen gas will be excited via an alternating high-voltage electrical current. Observe this light using your spectroscope. For Hydrogen, you should see 3 or 4 distinct lines, at 410 nm, 434 nm, 486 nm, and 656 nm. Note: the violet line at 410nm may not be visible to all people. Record the position of the emitted light directly on your spectral paper.

   In the space below, record the positions of the visible Hydrogen lines and relative intensity of each line. Label the wavelengths.

Gas: **Hydrogen**

V   I   B   G   Y   O   R

Now create a spectral library. View four more sources and draw their visible spectral lines below, using the Hydrogen lines as a reference. Record the wavelengths below the lines.

Gas: _____

V	I	B	G	Y	O	R

Gas: _____

V	I	B	G	Y	O	R

Gas: _____

V	I	B	G	Y	O	R

Gas: _____

V	I	B	G	Y	O	R

Now look at the Unknown Gas Combination. Draw the spectra you see. What two elements seem to best fit this combination?

V	I	B	G	Y	O	R

Continuum Spectrum

During this part of the lab, you will sometimes need to examine light from the sun using your spectroscope. **Do not look directly at the sun! This can cause permanent damage to your eyes.**

1. Use your spectroscope to investigate the spectra of each flashlight (both incandescent and LED). If it is sunny outside, you can also check the solar spectrum by looking at the reflection of the sun off of a piece of glass or a windshield. Do not look directly at the sun itself. How do the sunlight's spectra compare with the emission spectra?

2. How do the incandescent light bulbs compare with the LED from your cell phone flashlight?

3. Try looking at various sources of "white" light. White light is usually described as a combination of all of the other colors. Is this strictly true? Write down what you looked at and what colors you saw being emitted.

Absorption Spectra

1. Use your spectroscope to investigate the light emitted by the sun. **Do not point your spectroscope toward the sun at any point.** Instead, look at the reflection of the sun off the windshield of an automobile or look at the spectrum of the clouds. Do you see any absorption lines? Record your observations.

2. Look up a "Solar Spectrum" online. How do astronomers know what elements are found in the atmosphere of the sun?

3. Compare the Sun's spectrum with other stellar spectra online. Are they the same? Are they different?

Classifying the Spectra of Stars

So far we've been exploring the types of spectra observed here on Earth, but how does this apply to stars? We will use the VIREO simulation to examine some stellar spectra the way that astronomers observe them.

Instructions

1. Using a computer, open the VIREO simulation.

2. Click on "File" in the top menu bar and select Login. You can enter your names if you like, but it isn't necessary. Click on "OK" when you are done.

3. Once it displays "The Virtual Educational Observatory," click again on File and go to "Run Exercise." Then select "Classification of Stellar Spectra."

4. This simulation mimics running a telescope remotely. The first thing we need to do is select the telescope we want to use. Click on "Telescopes" in the menu bar and use the Optical 0.4 meter telescope, or you can try to "Request Time" on a larger telescope. A larger scope doesn't give you better data but does give it to you faster.

5. Once you have accessed the telescope and are in control of it, use the Dome toggle to Open the dome.

6. Once the Dome is open, click on the "Off" button below the Telescope Control Panel to access it. You will see the view from the wide field Finder scope of the telescope. The stars will be moving slowly across the field, so click on Tracking on the left side of the screen to have the telescope move with the stars. In the top right hand corner is the View box. Click on Telescope to focus on the view from the main telescope.

7. You should see two vertical red lines. Those represent the slit of the telescope's spectrometer, just like the slit in the handheld spectrometers. You want to use the NESW (North, East, South, West) buttons to align a star between the two lines. Make sure that you don't have two stars near each other, which will give you a mixture of theit two spectra. It should look something like this screenshot (figure 5.3)

8. Once you have your star aligned, click the "Access" button in the Instrument box to bring up the Spectrometer view. Click on "Go" and you will start collecting light. You want to wait for the Signal to Noise Ratio to be higher than 10 to make sure you have a clean signal. If your star is dim, this could take a while.

9. Once your signal is taken, click on File, Data, and Save Spectrum so we can analyze it.

10. Once you have saved your spectrum, close the Spectrometer window and go back to the VIREO Exercise window which should just be a mostly black box. Click on Tools and Spectral Classification.

11. To load your spectrum, click on File, Unknown Spectra, Saved Spectra, then click on your saved data. You should see a squiggly line with some deep troughs.

12. Click on File, Display, Comb. (Grayscale + Trace) to see a "photo" of the spectrum that will look more like what you saw with the handheld spectroscopes. Change the Display back to the Trace only for the next step.

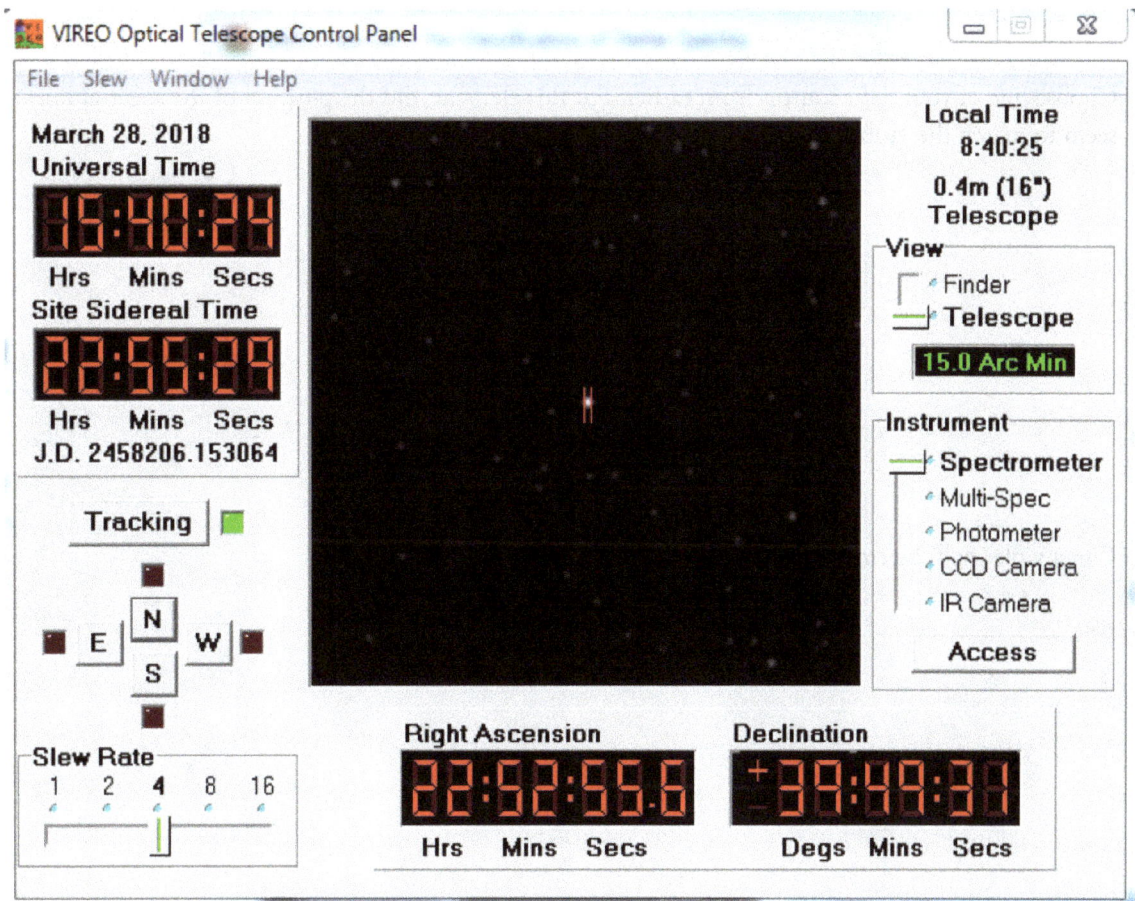

Figure 5.3

13. We need to compare this spectrum with some known spectra of stars. Click on File, Atlas of Standard Spectra, and select Main Sequence. You should see a selectable list of spectra in the top right hand corner of the window.

14. Go through the known spectra until you find one that is closest to your Unknown spectrum. Try to find a spectrum that matches your Unknown.

15. Once you have a close spectrum, you can check it by clicking on File, Display, Show Difference (Std – Unkn). This subtracts the two spectra and plots the result. If the spectra are nearly identical, the Difference should be a nearly flat line.

16. Record your spectra classification below. If the Sun is a G2 Main Sequence Star, is your star hotter or colder? Bigger or smaller?

Conclusions

1. Choose one of your light sources from earlier. For this element, did the position of the spectral lines seem to match the visible color of the light?

2. Can a violet light be emitting any orange or red wavelengths? Explain.

3. If I showed you a spectral pattern which matched one in your library, what could you tell me about the light which caused it?

4. Why didn't your unknown spectrum exactly match the standard spectrum? Astronomers will often refer to a star's spectrum as its "fingerprint." Why do you think this is?

Name _____ Section _____ Date _____

Group Members _____

Lab 6: **Hertzsprung-Russell Diagrams**

Purpose

To explore the relationship between luminosity and temperature for stars and to see how stars are only found in certain regions of the H-R diagram.

Background

The Hertzsprung Russell (H-R) Diagram is used to show where compare to other stars in terms of luminosity and spectral class. Luminosity refers to the absolute brightness of a star, and spectral class refers to the star's temperature and color. Astronomers organize stars on an H-R diagram by graphing temperature (spectral class) on a horizontal axis from highest (left) to lowest temperature (right). Luminosity is graphed on a vertical axis from dimmest (bottom) to brightest (top).

Instructions

Your task is to create an H-R diagram with an appropriate scale using the attached stellar data. Make sure to plot the Sun's position on the graph. You will then label groups of stars as shown on the example H-R diagram in your textbook. Remember to label all axes with appropriate units. As you label all the stars, use different symbols or colors for each group (25 brightest vs. 25 closest).

Questions

1. Which stars would be grouped together in the lower left-hand corner of the diagram? What are some of their properties (you may need to look in your book)? Label the graph.

2. Where does the "main sequence" appear on the H-R diagram? What does it represent?

3. Label the point on the diagram where the Sun appears. Show the Sun's path as it evolves off the main sequence (in essence — where is the sun going in the future)?

4. Where do the red giants appear on the H-R diagram (label the area)?

5. What are the stars in the lower right of the diagram? What are their properties?

6. Why are so many of the nearest stars not on the brightest star list?

7. Where on the diagram would you find protostars?

8. Use a computer to access the blackbody simulation at http://phet.colorado.edu/en/simulation/blackbody-spectrum. Use the temperature ranges to create "color bands" for each spectral type. You can either describe their colors or use colored pencils.

9. Pick a star in each spectral category (BAFGKM) and locate it in Stellarium. In the right-hand corner of the screen is the information about the star. Make a table of the star, its spectral type, its blackbody color, and the perceived Stellarium color.

10. Go to the following web page: http://astro.unl.edu/naap/hr/animations/hrExplorer.html. It requires Flash, so you may need to use one of the lab computers to access it. This page gives the size comparison of a star with the properties of the x on the H-R diagram. As you move the x up and down but keep the temperature the same, what happens to the comparison star?

11. As you move the x left and right but keep the Luminosity the same, what happens to the star?

12. Click the "Show Luminosity classes" tab under Options. Why are some stars "giants" and others are "dwarfs?" How do the blue stars compare to the red stars (besides being different colors)?

13. Under Plotted Stars, compare the nearest stars, brightest stars, and both distributions. How do they compare with your own H-R diagram?

14. What kind of star is the most common and closest star?

15. What kind of star is the most luminous?

16. What kind of stars are in the overlap between brightest and nearest stars?

Plotting Pulsating Variable Stars on an H-R Diagram

The table below contains variable stars that change both their luminosity and their spectral class. These stars are rare but extremely useful for measuring distances between clusters of stars and even other galaxies.

To show the entire cycle of change for variable stars, it is necessary to plot them twice — at maxima and minima. The spectral class column gives the maximum (left) and minimum (right), and the absolute magnitude (M_v) column gives the range from maximum (left) to minimum (right).

1. Plot each star at both maximum and minimum absolute magnitude along with the corresponding spectra class. For example, RT Aur will be plotted at (F4, –3.4), and again at (G1, –2.6). Use a different symbol from the other stars (such as a box, open circle, or triangle).

2. Draw a line connecting the two points.

3. Identify the type of each variable star by locating its position on the H-R diagram. Write it next to the star in the table below.

Table 1: Variable stars

Star	Type	Distance (Parsecs)	Spectral Class	Absolute Magnitude
RT Aur		480	F4 to G1	–3.4 to –2.6
Delta Cep		300	F5 to G1	–3.9 to –3.0
Rho Cas		3600	F8 to K0	–8.7 to –6.6
T Cas		1700	M6 to M9	–3.2 to +0.8
TU cas		1100	F3 to F5	–3.3 to –2.0
X Cyg		608	F7 to G8	–3.3 to –2.3
T Cep		210	M5 to M8	–0.6 to +3.7
Y Oph		880	F8 to G3	–3.8 to –3.3
RS Boo		1300	A7 to F5	–0.9 to +0.2
VX Her		2100	A4 to F4	–1.7 to –0.4

4. Are the variable stars all the same spectral type of star? Are they the same evolutionary type of star?

Star Name	Label	Brightness (apparent magnitude)	Luminosity (absolute magnitude)	Distance In parsecs	Spectral Class
Polaris	αUMi	2.0	−4.5	200	F8
Mira	oCet	2.0	−1.0	40	M6
Algol	βPer	2.1	−0.5	31	B8
Aldebaran	αTau	0.8	−0.8	21	K5
Capella	αAur	0.1	−0.6	14	G8
Rigel	βOri	0.1	−7.0	270	B8
Bellatrix	γOri	1.6	−4.1	140	B2
Mintaka	δOri	2.2	−6.0	450	B0
Betelgeuse	αOri	0.4	−5.9	180	M2
Sirius	αCMa	−1.4	1.4	2.7	A1
Castor	αGem	1.6	0.8	14	A1
Procyon	αCMi	0.4	2.7	3.5	F5
Pollux	βGem	1.2	1.0	10.7	K0
Regulus	αLeo	1.3	−0.8	26	B7
Merak	βUMa	2.4	0.6	23	A1
Dubhe	αUMi	1.8	−0.6	30	G9
Denebola	βLeo	2.1	1.6	13	A3
Mizar	ζUMa	2.1	0.0	26	A2
Spica	αVir	1.0	−3.1	65	B1
Arcturus	αBoo	−0.1	−0.2	11	K1
Antares	αSco	0.9	−4.7	180	M1
Vega	αLyr	0.0	0.5	8.1	A0
Altair	αAql	0.8	2.2	5	A7
Denebola	αCyg	1.2	−7.3	500	A2
Markab	αPeg	2.5	0.0	32	B9

Table 3: 25 of the 100 nearest stars

Star Name	Label	Brightness (apparent magnitude)	Luminosity (absolute magnitude)	Distance In parsecs	Spectral Class
15 1915 A	N1	8.9	11.2	3.5	M4
15 1915 B	N2	9.7	12.0	3.5	M5
L347-14	N3	13.7	14.8	5.9	M7
symbol oDra	N4	4.7	5.9	5.7	K0
Altair	N5	0.8	2.2	5	A7
δPav	N6	3.6	4.7	5.8	G7
HR 7703 A	N7	5.3	6.5	5.8	K4
HR 7703 B	N8	11.5	12.7	5.8	M5
−45 13677	N9	8.0	9.0	6.3	M0
61 Cyg	N10	5.2	7.5	3.4	M5
−39 14192	N11	6.7	8.7	3.9	M0
−49 13515	N12	8.9	10.6	4.6	M3
εInd	N13	4.7	7.0	3.5	K5
DO Cep A	N14	9.8	11.8	4	M3
DO Cep B	N15	11.4	13.4	4	M4
L789-6	N16	12.6	14.9	3.4	M6
−21 6267 A	N17	9.3	11.0	4.6	M2
−21 6267 B	N18	11.0	12.7	4.6	M3
43 4305	N19	10.0	11.5	5	M5
Ross 780	N20	10.2	11.7	4.9	M5
−36 15693	N21	7.4	9.6	3.7	M2
56 2966	N22	5.6	6.5	6.6	K3
Ross 246	N23	12.2	14.8	3.2	M6
1 4774	N24	8.9	10.0	6.1	M2
τCeti	N25	3.5	5.7	3.6	G8

Lab 7: **Physical Characteristics of Stars**

Purpose

To learn the relationships between:
- A star's mass and its luminosity
- A star's mass and its life expectancy
- A star's mass and its temperature

Materials

▷ Notebook paper

▷ Computer

Key Ideas

You will graph relationships between some properties of stars, and use these graphs to answer questions about the stars. You will also make predictions about unknown stars.

Background

The single most important physical property of a star is its mass. Most of its other properties depend, at least in part, on mass.

Some properties of a star:
- **Luminosity** is the total radiant energy output per unit time. It is usually expressed in ergs per second. The sun's luminosity is $3.83*10^{33}$ ergs/sec of which $2*10^{24}$ ergs/sec fall on Earth.
- The **effective temperature (T_{eff})** is approximately the surface temperature of a star. The T_{eff} of our sun is 5800K.
- The **life expectancy** of a star is the length of time it exists as a stable, main-sequence star. Our sun has a life expectancy of approximately 10 billion years. Life expectancies are approximated through observation and knowledge of nuclear physics.

The Data

Table 1 shows average properties for main sequence stars by spectral type (S_p), it's absolute magnitude (M_v), it's mass in terms of a solar mass (M/M_o), it's radius in terms of a solar radius (R/R_o), and its luminosity ratio with respect to the solar luminosity (L/L_o), the effective surface temperature (T_{eff}), and the star's lifetime (Y) in millions of years.

Table 1. Average Properties for main sequence stars by spectral type.

Sp	Mv	L/LSun	M/MSun	R/RSun	Teff(K)	Y(Myr)
O5	−6.0	22000	40	18.0	50000	0.1
B0	−4.1	3800	17	7.6	27000	
B5	−1.1	240	7	4.0	16000	80
A0	+0.6	50	3.6	2.6	10400	
A5	+2.1	13	2.2	1.8	8200	2000
F0	+2.6	7.9	1.8	1.3	7200	
F5	+3.4	3.8	1.4	1.2	6700	5000
G0	+4.4	1.5	1.1	1.04	6000	
G5	+5.2	0.72	0.9	0.93	5500	10000
K0	+5.9	0.38	0.8	0.85	5100	
K5	+8.0	0.055	0.7	0.74	4300	20000
M0	+9.2	0.018	0.5	0.63	3700	
M5	+12.3	0.0011	0.2	0.32	3000	50000

Source: Zeilik, M., and E.v.P. Smith, *Introductory Astronomy and Astrophysics*, Saunders College Publishing, 1987.

Part I: Graph the Data

This part of the lab requires you to find several comparisons between parts of the data in Table 1.

1. Find the relationship between luminosity (L/L$_{Sun}$) and mass (M/M$_{Sun}$).
    a. To do this, Graph the data using Excel. Make the x-axis M/M$_{Sun}$ and the y-axis L/L$_{Sun}$.
    b. Draw a best-fit power line through the plotted points by selecting Chart – Add Trendline.
    c. Under the Trendline options, check the "show equation on chart" box. From the equation, find the slope of the trendline. Call the slope this symbol: α.
2. Repeat this process comparing Temperature and Mass. (Mass is still on x-axis.)
3. Repeat this process comparing Life-Expectancy and Mass.

Part II: Summary Questions

Please produce a report that includes the graphs produced above and a discussion of the following questions:

1. Describe the relationship between a star's luminosity and its mass. How is this significant?

2. In 1924, Sir Arthur Eddington derived the mass-luminosity relation and found it to be $L/L_o = (M/M_{Sun})^\alpha$, where $\alpha \cong 3.5$. How does your relation compare?

3. Describe the relationship between a star's life expectancy and its mass. Why is this important?

4. Describe the relationship between a star's temperature and its luminosity. Why is this important?

5. Why does a low mass star live longer than a high mass star?

Part III: Predictions

Use your mathematical relationships, graphs, and the data in Table 1 to predict the Luminosity, Life-expectancy, and Effective temperature of each of the following average, stable, main-sequence stars. Add the resulting table to your report.

1. Star A is one-tenth the mass of the Sun
2. Star B is three times the mass of the Sun
3. Star C is five times the mass of the Sun
4. Star D is twenty-five times the mass of the Sun
5. Star E is our Sun

Table 2: Predicted Stellar Properties

Star	M/Mo	L/Lo	Y(Myr)	Teff(K)
A	1/10			
B	3			
C	5			
D	25			
E	1			

Part IV: Observations and Inferences

1. In this last part, you are going to use inferences from your data to create testable statements. First, looking at the data you have made, what stars do you think would be the brightest in the sky?

2. From your data, what stars do you think would be the most common? Why do you think this?

3. We will use Stellarium to make a limited survey of these. First look for 5 bright stars that have names. Click on them and record their name, their RA and Dec, and their spectral type.

4. How many O and B stars were in your sample of stars? If you didn't find any, that's ok, we'll explore that more in the next question.

5. Find the "List of Brightest Stars" on Wikipedia. How many O stars are there on the list? How many B stars?

6. Pick two O stars and two B stars that have names on the list. Observe them in Stellarium. Are they particularly bright? Why or why not?

7. Why are O and B stars so rare, using your observations and your data?

8. From your data, what should be the most common type of star?

9. Using the brightest stars list again, are there any main sequence M stars? Note: Main sequence stars should have a V next to their spectral type letter.

10. If the universe is roughly 14 billion years old, how does that compare with the lifetime of M class main sequence stars?

11. Should M class main sequence stars be common or uncommon? Why don't we see them very easily?

12. What two factors affect how bright a star appears in the night sky?

Lab 8: **The Milky Way Galaxy**

Purpose

To examine the motion of stars compared to the motion of spiral arm structures in the Milky Way galaxy. To examine the observations that cause astronomers to conclude that dark matter exists.

Materials

▷ Computer

▷ Graph paper

▷ Ruler

▷ Compass

The Milky Way Galaxy

1. How do we see the Milky Way Galaxy? Search the internet for images of the Milky. What do the images you found all have in common?

2. Look up the Gaia mission and its map of the Milky Way galaxy at http://sci.esa.int/gaia. Does it look different from other maps of the Milky Way?

3. Do some online research on the Gaia mission. Why are astronomers so interested in this map? What does it have direct measurements of?

21-cm Hydrogen Line

In the visible part of the spectrum, light sources are predominately stars which are hot enough to emit in those visible wavelengths. However, the bulk of a galaxy is cool, neutral hydrogen. In fact, the most common element in the entire universe is hydrogen. That hydrogen also emits light, but in the microwave where we can't see it. Molecular gas collects in the spiral arms of the galaxy (due to density waves), so most of the 21-cm wavelength emission comes from those spiral arms.

First, let's look at the way light moves in these spiral arms. On figure 8.1, draw 5 lines of sight starting from the position of the Sun so that we get identifiable clouds at a variety of distances. These lines of sight

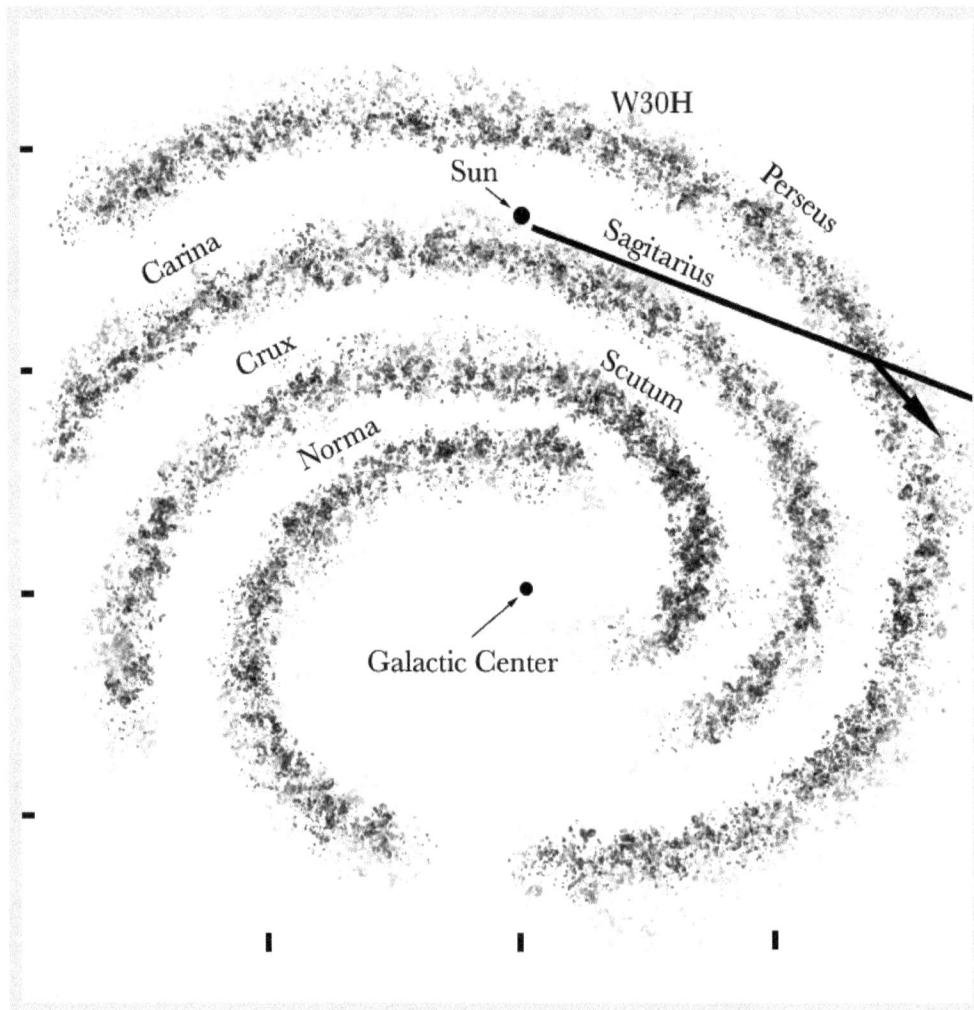

Figure 8.1

should connect with the spiral arms and avoid the galactic core. Draw a dot where the lines cross the spiral arms and then add an arrow for what direction the dot would be moving. An example line has been added to figure 8.1.

$$v_r = c * \left(\frac{\lambda_0}{\lambda_{obs}} - 1\right)$$

Doppler Effect

Now that we've determined which direction the light moves, the second step we need is a way to measure its velocity. Fortunately, the Doppler Effect works for light in the same way that it works for sound. If a molecular cloud is approaching the Earth along our line of sight, it's 21-cm emission will be shifted to shorter wavelengths. If it is moving away from the Earth along our line of sight, it will have a longer wavelength. In the equation below, c is the speed of light, and we want to use 300000 km/s.

Fill out Table 1 to determine the radial velocities of the following clouds. Note: A spreadsheet can help you out a great deal here.

Table 1

Object Observed	Distance from Galactic Center (kpc)	Lamda observed	Rotational Speed
A	0.2	20.993	
B	1	20.9825	
C	3	20.98425	
D	5	20.986	
E	7	20.98425	
F	9	20.9846	
G	11	20.986	
H	13	20.98425	
I	15	20.98425	
J	17	20.98425	
K	19	20.9839	
L	21	20.9839	
M	23	20.98355	
N	25	20.98355	
O	27	20.9832	
P	29	20.9832	
Q	31	20.98285	
R	33	20.98285	

Galaxy Curve Vs. Gravitational force vs. Solid Body Rotation

Now that we have velocities, we can create our Rotation plots. Using Standard Candles and other distance tools, we can find the distance between each cloud and the Galactic Center.

Solid Body Rotation

The stars in the nucleus seem to be following solid body rotation. In other words, the farther out they are, the faster they are rotating. Create a plot for what the galaxy would be like with solid body rotation by using the following equation:

$V = 100 * D$ Where V is velocity and D is distance.

Keplerian Rotation

It would make sense that stars orbit the Milky Way the same way the planets orbit the Sun. In this way, the farther away stars are from the Galactic center, the slower they should be orbiting it. Use the following equation to model Keplerian rotation:

$V = 100 / D^2$

Plot your rotation curve, the solid body rotation model, and the Keplerian rotation all on the same graph. How does the observed rotation curve compare with these models?

Modeling Dark Matter and Galaxy Rotation

Using computers or laptops in the lab, download the Galaxy Rotation java simulation from Blackboard. Use the following suggested activities to explore the simulation.

1. Differential Rotation

In the Galaxy Rotation window of the simulation, you can see the core of the galaxy (blue circle at the center) and the initial positions of a set of smaller red circles representing the galaxy stars (visible matter). If you press the play button, the simulation begins and stars rotate. The rotation occurs in a plane and along circular trajectories with angular and linear velocities which are a function of the distance from the center. Since angular velocity is not the same for all galaxies, the shape of the distribution of stars changes, and after a while, it shows the usual spiral structure.

Although differential rotation is not the only cause of the observed spiral structure, it is quite interesting to show an important physical effect: if we allow the simulation to run long enough, we can observe that the spiral structure is not stable but that it disappears, with stars rotating around the central core in a random distribution.

1. Do we observe the spiral arms of the Milky Way getting more tightly wound? What does this mean about how good our simulation is?

2. Visible and Dark Mass

Reset the simulation and press the play button. Stars (red circles) begin to rotate. The value of the rotation velocity is close to the value observed for the spiral galaxy NGC2403. Check the "show plots" option, and a new window will pop up. There you can see the plots of both the rotation curve of the galaxy (upper plot) and the distribution of the total mass within the galaxy (lower plot) using Newtonian dynamics.

Note: at large distances from the galaxy center, the rotation velocity is almost constant, which means that the total mass is directly proportional to the distance from the center and therefore still growing. In other words, we have not yet reached the "edge" of the galaxy at the distances we have reached with our measurements.

By checking the option "show dark m. panel," you can see a new window where the first line displays the value of the ratio of visible to total mass of the galaxy (within the region covered by the observed stars). It is quite remarkable that you can see only 12% of the total gravitating mass. We can deduce that within the region we display there must be a large amount of matter that we do not observe since it does not emit any kind of radiation and that we call "dark". Then, in the Dark Matter Panel we can check the option "show no DM galaxy" in the main Galaxy Rotation window a set of yellow circles appears.

These circles represent the positions that stars would have if no dark matter were present. It is quite clear that in this case the galaxy appears, and rotates, in a very different way. This difference is also clear in the Plots window where yellow points indicate the distribution of rotation velocity and mass for the no dark matter case. The difference with the observed curves is really remarkable.

3. Where is the dark matter?

As a continuation of the previous activity, we can ask how dark matter is distributed. For example, it could be attached to individual stars. In order to check this hypothesis, you can adjust the value of the "Star mass" parameter in the main Galaxy Rotation window. This parameter is initially set to 1, the observed value. Since the amount of the total mass is nearly 8 times the amount of mass in stars, you can set this parameter to 8. By doing so, you can see that the ratio of visible to total mass in the DM panel rises to 1, but the distribution of yellow circles (the no dark matter stars) is very different from observed stars. You will find that it is not possible to reconstruct the observed galaxy rotation curve using only the mass of stars!

1. In the end, what can we conclude about the distribution of dark matter?

4. Interpretation of the Galaxy Rotation Curve

Now you can follow up activities 2) and 3) by introducing the presence of a certain amount of dark matter distributed within the entire galaxy. In order to do this, check the "show DM contribution" box in the Dark Matter panel. In the Plots window, new curves appear. In the upper plot, you can see the rotation curve (blue symbols) that is determined by adding the contributions of visible and dark matter at each distance from the center. In the lower plot, you can see two new curves: one is the distribution of dark matter, and the other is the distribution of visible plus dark matter. This last curve is quite close to what is actually observed.

Also in the main Galaxy Rotation window, a new set of purple symbols appear. These represent the positions of stars taking into account both visible and dark matter. The density of dark matter decreases from the center to the outer region of the galaxy. The length scale of the dark matter controls the decrease of the dark matter density as a function of the distance from the center. You can check the option "show DM density" in the Dark Matter Panel in order to see a 3D plot of the dark matter density. It is possible to change this parameter to try to reach better agreement between both the observed and calculated rotation curves and the observed and calculated star distributions.

1.   What can you conclude about dark matter's contribution to correctly interpreting the rotation curve of spiral galaxies?

Ramifications

1.   There is only one known force in the universe that can control and constrain stars in a galaxy: Gravity. As objects get further away from what they orbit, the force of gravity decreases as the distance squared. For the rotation curve of the Milky Way to be flat, what must be happening to the stars out in the disk? What are some potential explanations for what is going on?

2. For the nucleus of the Milky Way, the stars seem to be orbiting like a solid body. What does this tell us about the mass density of the nucleus?

3. Look up "dark matter" on NASA.gov and Wikipedia. What is dark matter? What is NASA trying to learn about it?

4. Look up information on the recently discovered galaxy that doesn't have any dark matter. How was that determined? What are some of the ramifications of this discovery?

Name _____ Section _____ Date _____

Group Members _____

Lab 9: **Variable Stars and Galaxies**

Purpose

- Find a relationship between period and luminosity for Cepheids
- Use this relationship to estimate distances to galaxies
- Interpret graphs
- Look at different types of galaxies

Materials

▷ Computer

▷ Graph paper

Key Ideas

You will learn how astronomers measure distances to far away stars using Cepheid Variable stars. You will also examine different types of galaxies.

Background

Our goal is to measure distances to stars and distant galaxies. Parallax can be used for nearby stars within a few hundred parsecs of us. This encompasses the closest few thousand stars, in our immediate neighborhood in the Milky Way. But to determine larger distances, astronomers use **variable stars**, whose luminosities (and therefore apparent magnitudes) vary with time. There are five kinds of stars used to find interstellar distances:

1. Novae
2. Cepheids
3. RR Lyrae stars
4. Supergiants
5. Eclipsing binaries

You will study the nature of **Cepheid Variables** in this lab activity. Cepheids are giant stars, with luminosity between 800 and 10,000 times that of our sun. Cepheids change luminosity in a periodic cycle. This bright-dim-bright period typically takes between 1 and 50 days.

Cepheids are important to astronomers because there is a direct relationship between a Cepheid's period and its average luminosity. The longer the period, the brighter the star. This was first noticed by Henrietta Leavitt, of Harvard College Observatory, and reported in 1912. She discovered over 1,700 variable stars, and found a correlation between apparent magnitude and the period in Cepheid variables in the Small Magellanic Cloud (SMC).

The apparent magnitudes of the Cepheids in the SMC are around +15, much dimmer than those Cepheids nearby in our own galaxy. This indicated that the SMC must be rather far away. By finding a Cepheid variable near enough to the Earth that its distance could be measured by parallax, astronomers could use that star to relate the absolute magnitude to apparent magnitude. This allows us to use the Leavitt relationship to measure distance to any objects containing Cepheid stars, including the SMC.

Part 1: Period – Luminosity Relationship for Cepheids

Table 1 lists values of period (log P) and average apparent magnitude (M_v) for several Cepheids. Since these Cepheids are all in the same cluster, their distances from Earth are approximately the same.

Table 1—Cepheids in the Small Magellanic Cloud

HV	Log P	mv	HV	Log P	mv
837	1.63	13.2	1945	0.81	15.2
840	1.52	13.4	1954	1.22	13.8
844	0.35	16.3	1967	1.45	13.5
847	1.44	13.8	1987	0.50	16.0
1809	0.45	16.1	2019	0.21	16.8
1825	0.63	15.6	2035	0.30	16.7
1837	1.63	13.1	2046	0.41	16.0
1873	1.11	14.7	2060	1.01	14.3
1877	1.70	13.1	2063	1.06	14.5
1903	0.71	15.6	11182	1.60	13.8

1.   Plot these data with log P on the horizontal axis and m_v on the vertical axis using a spreadsheet.

2. Locate HV 837 and HV 844 on your graph and put an x through each. Use the graph to answer these two questions:

   a. Which of the two stars takes longer to complete one cycle?

   b. Which of the two stars has the greater number for its magnitude?

3. Which star has the greater luminosity, HV 837 or HV 844? Explain how you can tell.

4. Use a computer find a **linear best-fit** through the plotted points. This line represents the period-luminosity relation. Write the equation from your best fit line below.

5. Imagine that another galaxy contains a Cepheid with the same period as HV 837. If the new Cepheid appears dimmer than HV 837, which galaxy is closer? Explain how you can tell.

Part 2: Find the Period of a Cepheid

Table 2 gives light curve data for Cepheid Variable HV 843. Remember that brightness is an inverse relationship to the magnitude number.

Table 2. Light curve data for HV 843

Day	mv		Day	mv		Day	mv
0	15.31		16	15.18		32	14.93
1	15.22		17	15.05		33	14.73
2	15.1		18	14.91		34	14.37
3	14.92		19	14.60		35	14.40
4	14.70		20	14.37		36	14.50
5	14.35		21	14.42		37	14.61
6	14.37		22	14.52		38	14.72
7	14.45		23	14.62		39	14.87
8	14.56		24	14.74		40	15.00
9	14.58		25	14.87		41	15.13
10	14.79		26	15.04		42	15.22
11	14.93		27	15.18		43	15.32
12	15.07		28	15.29		44	15.24
13	15.20		29	15.30		45	15.14
14	15.31		30	15.20		46	15.00
15	15.28		31	15.07		47	14.79

1. Plot this curve in a spreadsheet. Do not add a trendline!

2. Using your plot, estimate the period of the star. (The period is the number of days from peak to peak in one cycle.)

    P =

3. Take the logarithm of this period.
    Log P =

4. From your period-luminosity graph in part 1, estimate the magnitude of this star.
    m_V =

5. Use your period graph, to determine the maximum and minimum magnitudes of HV 843.
    m_{max} =

    m_{min} =

6. Find the apparent magnitude m_v, which is the average of m_{max} and m_{min}.

7. How does this value compare to your estimate in #4?

8. Now that you have the log P of the variable star, you can find the distance by comparing the absolute magnitude M_V and the apparent magnitude m_V. To find M_V, use the following equation:
    M_V = -2.43*(Log P -1) – 4.05

9. Now that you have M_V, you can use the **distance modulus** to find the distance to the variable star. Use the following equation to find the distance in parsecs.
    $d = 10^{(0.2(m_V - M_V + 5))}$

Questions

1. Describe the general relationship between the period of light variation and the luminosity of a Cepheid variable star.

2. Describe how the Cepheid Variables are used to calculate interstellar distances.

Part 4: Galaxy Zoo

1. Describe the three main classes of galaxies: Spiral, Elliptical, and Irregular.

2. Go to the Galaxy Zoo (http://www.galaxyzoo.org/) and spend some time as a group classifying 10–12 galaxies. What categories do they use? Why do you think this is?

3. It's generally thought that irregular galaxies are created by collisions. What observational evidence is there that galaxies collide?

4. Watch the following YouTube video of galaxies colliding (https://youtu.be/C0XNyTp5brM). How well does this model seem to match observations?

Go to Wikipedia.com or Nasa.gov to find the following information:

5. When did the Hubble space telescope launch?

6. What issues did HST have when it first launched?

7. What is the Hubble Deep Field image? How long was the exposure? What is the angular size or dimensions of the image?

8. Estimate how many galaxies are visible in the image.

9. What is the Hubble Ultra Deep field image? How is it different than the Deep Field image? How long was the exposure?

10. What kinds of galaxies are observed in the Deep Field and Ultra Deep Field images?

11. Some astronomers think that there should be *Giga* Deep Field image where HST would be pointed on a blank spot on the sky for 500 days. Do you think this is a good idea or not? Why or why not?

Lab 10: **Expansion of the Universe**

Purpose

- Model an expanding universe
- Become familiar with some observations which suggest a Big Bang origin
- Discover current findings in cosmology

Materials

▷ Balloon

▷ Marker

▷ Computer

Key Ideas

Using a simple model, you will become familiar with the relation between receding galaxies and an expanding universe.

Background

Cosmology is the search for origins. It seems as if everyone wants to know how the Universe began. The Big Bang Theory is the result of several important observations. In 1927, Edwin Hubble observed that galaxies are red shifted, and moving farther and farther away from us. Second, he determined that the farther away a galaxy is from Earth, the faster it is receding.

If the universe is expanding, then one can assume that the galaxies that compose our universe were once much closer together than they are now. By simply measuring how far apart galaxies are and how fast they are moving, cosmologists determine what's called the Hubble Constant, which gives the recessional velocity of galaxies. This is much easier to measure than the current positions of galaxies. Remember, distances to galaxies are typically measured by finding Cepheid variable stars and observing their period.

Part I: Model of Expanding Universe

The universe has 4 dimensions: height, width, depth, and time. They are bound together as spacetime. As the universe expands, the view from any one place in the universe remains the same in the sense that the viewer has the same neighbors as before. In this activity, you are going to create a model of the expanding Universe.

1. One way to model the galaxy expansion is with a balloon. Take a deflated balloon and use a marker to draw some dots on it. The dots represent galaxies and the balloon represents spacetime.

2. Inflate the balloon about halfway, but don't tie it off. Look at how the dots have all gotten farther away from each other. Inflate the balloon even more. If you pick a dot as a "home galaxy," in what direction do all the other dots seem to move?

3. Next compare the pictures of Galaxy Field A and Galaxy Field B. How are they similar? How are they different? Try holding Field B behind Field A to the light and lining up two galaxies. Does it look the galaxies move between Field A and Field B?

4. Galaxy Field A represents that universe as it was 1 billion years ago. Galaxy Field B represents the universe as we see it presently. Which galaxies have moved between the past universe and the present universe?

5. To make life a bit easier, look at Labeled Galaxy Field 1 and 2. Choose one galaxy to be the "home" galaxy. A home galaxy doesn't seem to move because we are in it, so that will be a measuring point.

6. From your home galaxy, measure the distance to 10 other galaxies in **Labeled Galaxy Field 1** and record their names and the distances below.

Galaxy Name										
Distance										

7. Repeat your measurements from your home galaxy to the same galaxies but in **Labeled Galaxy Field 2**.

Name										
Distance										

8. Using Google Sheets or Excel, make a plot where the horizontal axis is the distance to each galaxy from Galaxy Field 1 and the vertical axis is the **difference** between the distances in Galaxy Field 2 and Galaxy Field 1.

9. This graph represents how "fast" the galaxies are moving relative to their initial distance. How does this compare to Hubble's Law?

10. Does this help explain why redshifts are observed in all galaxies outside of the Local Group?

Part II: The Big Bang

If we run the expansion process backward, we get two results. The first result is that it took approximately 14 billion years for the Universe to grow to its present size. Second, an awesome event must have caused the galaxies to go flying away from one another — the Big Bang. In addition to the predictions from Einstein's General Theory of Relativity, here are four observations and inferences that suggest that a *Big Bang* did actually occur very long ago.

Observation	Inference
Almost all galaxies are red-shifted	Almost all galaxies are moving away from the Milky Way
The most distant galaxies exhibit the greatest red-shift	The most distant galaxies are moving away the fastest
Recessional velocity is proportional to distance from us (The value is called the Hubble Constant.)	The Universe has been expanding for 8 to 15 billion years.
The Cosmic Background Explorer (COBE) found that the temperature of intergalactic space was not zero	The universe has not yet cooled from the rapid Big Bang expansion.

Other than those listed, give two examples of information from astronomy that demonstrate the difference between an observation and an inference.

Observation	Inference

Part III: Drawing parallels: the fifth Dimension

1. First, play a game of 2D Tic Tac Toe, then play a game of 3D Tic Tac Toe.

   2D: http://www.twoplayergames.org/play/845-Tic_Tac_Toe_Paper_Note.html
   3D: https://www.mathsisfun.com/games/foursight-3d-tic-tac-toe.html

2. How much more complex is 3D versus 2D?

3. The connection between 2D and 3D is this: Imagine a beach ball that is casting a shadow on the ground. The shadow is in the shape of a circle. Thus, a circle is a 2D projection of a sphere which is 3D. How could you take a cube and create a square? What other 2D shapes have a 3D version?

4. This idea is expanded by the concept of "hyper" shapes. Search the internet for pictures of hyper shapes such as a hypercube, hypersphere, or just hypershapes. What do they share in common with 3D shapes? What is different?

5. One way to draw a cube is to draw two squares and then connect the corners with lines. Draw two cubes near each other below. Try to make a hypercube by connecting the corners of the two cubes together with lines.

6. Now imagine the universe is a hypersphere. How is this like the dots on the balloons? How might we be able to "see" the expansion of the universe this way?

7. Download a 4D Rubik's Cube here: http://www.superliminal.com/cube/cube.htm. How much more complex is it than a standard cube? Why does it seem to float? Could such an object exist outside of a computer?

www.ingramcontent.com/pod-product-compliance
Lightning Source LLC
Jackson TN
JSHW051102050325
79772JS00004B/7